The Miller
in Eighteenth-Century Virginia

THE READER of this account, being of open mind and charitable disposition, as good men and women have ever been, will readily recognize that whatever may appear in these pages to the discredit of millers in times past, cannot be taken to reflect in any fashion upon the present master of Mr. Robertson's windmill. Indeed, the age-old repute of the calling is as distasteful to him and his colleagues of today as it would be inappropriate if applied to them.

Unhappily, it cannot be denied that millers of an earlier day—those of Chaucer's generation, for example—left something to be desired in the way of scruple. That gifted story-teller and honest reporter of the age in which he lived gave prominent place in his *Canterbury Tales* to two millers. One of these was the villain and ultimate victim in the Reeve's Tale: "A thief he was, forsooth, of corn and meal; And sly at that, accustomed well to steal."

The other miller of the *Canterbury Tales* was himself one of the pilgrims, as merry and uncouth a rogue as one could find in any band of cathedral-bound penitents: "He could steal corn and full thrice charge his tolls; and yet he had a thumb of gold, begad." That last remark, an allusion to the proverb that "Every honest miller has a thumb of gold," cut a broad swath indeed. Only Chaucer's own regard for truth could have moved him thus to dignify the popular belief that among millers integrity was as rare as twenty-four-carat thumbs.

Similar distrust can be discerned in early feudal and manorial laws in England which prescribed certain methods of operation for grist millers and established corresponding penalties for violation. The miller was directed to charge specified tolls for his services, and no more. The lord of the manor got his grain ground "hopper free," since he generally owned the mill and held the local milling monopoly. Under the thirteenth-century Statute of Bakers, chartered landholders paid the miller one-twentieth of the grain he ground for them, tenants-at-will gave one-sixteenth, while bondsmen and laborers had to part with one-twelfth of what they brought to the mill.

The same law also required that the miller's "toll-fat" (or dish) and "sceppum" (or scoop) used to measure grain be accurate. The manorial seal on a measure testified that it had been compared with the standard measure and found exact. But millers in all lands and times (the present excepted, of course) have been adept at finding ways to outwit law and customer at the same time.

A method popular among some millers was to build square housings for the millstones, thus providing four innocent corners in which quite a bit of meal could collect. The more artful members of the craft built a concealed spout that carried a small proportion of the meal to a private bin while the visible spout delivered the bulk of it to the customer's container.

Other stratagems, too varied and too numerous to list here, testify to the craftiness of many millers. The lengths to which the law went in trying to keep up can be seen in an English statute of 1648. This law, closing one loophole through which a miller could levy a hidden toll, allowed him to keep no hogs, ducks, or geese in the neighborhood of the mill, and no more than three hens and a cock.

All of this ingenuity, most of the popular suspicions of the

THE MILLER
in Eighteenth-Century
VIRGINIA

An Account of Mills &
the Craft of Milling, as well as
a Defcription of the Windmill
near the Palace in *Williamſburg*

Williamſburg Craft Series

WILLIAMSBURG
Publiſhed by *Colonial Williamſburg*
MMI

milling craft, and some of the legal restrictions consequent upon both, crossed the Atlantic along with the millers and millwrights who came to the colonies.   More details of this in a moment; meantime, what of the mill that belonged to the man that owned the name of rascal?

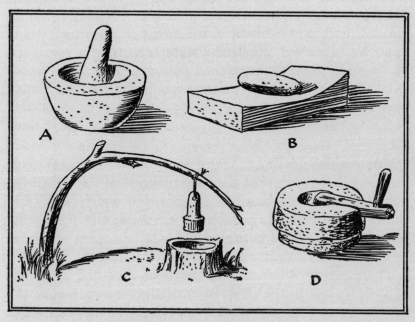

Some simple grist mills: (A) stone mortar and pestle; (B) saddlestone and metate; (C) sappling-and-stump type of mortar and pestle, often used by early colonists; (D) Roman quern.

THE OLD MILL STREAM OF HISTORY

For uncounted generations in every pre-mechanical civilization grain has been ground in a variety of one-woman-power devices.   Pounding with mortar and pestle was one of the earliest and is still the crudest of these devices.   The saddlestone-metate device, still to be seen in some areas of Central and South America, substituted a rolling, sliding motion to the upper stone that rubbed and sheared the grain. Finally, the Roman quern, rotating continually in the same direction and shearing the kernels between grooved faces of matched stones, opened the door to the use of natural instead of muscle power.

History does not record the name of the man, probably a Greek, who first harnessed natural power to grind grain between opposing stones. Possibly it happened when his wife handed him the family quern with the command, "Here, you do it!"—in Greek, of course. What he did, instead of earning his bread in the sweat of his brow, was to apply brain power. He fixed a water wheel to the lower end of a vertical shaft and attached the upper end to the upper stone of his handmill. And then, no doubt, he went fishing in the millstream while the flowing water did his work.

In the anonymous Greek's footsteps, a Roman named Vitruvius made the arrangement more flexible by introducing wooden gearing to transmit the power. Others made further improvements in the slow progress of time until the watermill was a reasonably efficient and widely used machine. The Domesday Book, or census of the year 1080, recorded 5,624 mills in England alone, all operated either by animal or water power.

The identity of the man or men who invented the windmill is also lost in the mists of antiquity—or at least of the Middle Ages. The earliest authenticated reference to a windmill in western Europe refers to one that stood in France about the year 1180. The next known reference dates from 1191 and concerns a windmill in England. Both were post mills, more or less like the reconstructed windmill of William Robertson in Williamsburg. This is the simplest among several types of wind-operated mills and was the type first adopted in Europe, in England generally, and in the colonies.

Anyone who has read much poetry cannot fail to realize that a watermill is by nature a more romantic machine than a windmill. Poets recognize this as a fact, and perhaps nonpoets who have spent some well remembered moments down by the old mill stream will agree. But this is not to say that windmills are lacking in emotional appeal and romantic inspiration. Far from it. As Robert Louis Stevenson wrote:

There are few merrier spectacles than that of many
windmills bickering together in a breeze over a woody
country, their halting alacrity of movement, their pleas-
ant business of making bread all day with uncouth ges-
ticulations, their air gigantically human, as of a creature
half alive, put a spirit of romance into the tame land-
scape.

*One of the earliest medieval illustrations of a windmill in England is this brass
plaque (here redrawn) in St. Margaret's Church, King's Lynn, Norfolk. It tells
the ancient joke on the good farmer riding to the mill: To relieve his tired horse
of the burden, he carried the sack of grain on his own back! Note that the mill was
a post mill and that it was slightly head sick.*

PUTTING THE WIND TO WORK

A post mill, as its name suggests, perches somewhat like
a flagpole-sitter on the end of a sturdy post held upright by
a timber framework. The Laws of Oleron, a breezily-worded
maritime code adopted in England about 1314, stated that
"Some windmills are altogether held above ground, and have
a high ladder; some have their foot in the ground, being, as
people say, well affixed." In the latter case the substruc-
ture of timber bracing was not above ground, but buried in
a mound of earth.

The comparison to a flagpole-sitter is perhaps misleading,
for the post does not end where the mill house begins.
Rather, it enters the body of the mill through a loose-fit-
ting collar beneath the lower floor and extends some half way
up into the mill, where it ends in a pivot bearing. The entire weight of the mill—sails, body, millstones, shafts, gears,
grain, meal, and miller (to say nothing of the mill cat, kit-
tens, and resident mice)—rests on this single bearing at the
top of the great post.

Keeping so much weight in stable balance was no great problem for the millwright as long as the mill did not move. The collar or ring bearing around the post kept the body from tipping far in any direction—or was supposed to. Moreover, the millwright estimated the weights of the various elements and positioned them appropriately about the pivot. Of course, things sometimes came out wrong. A mill that tipped incurably forward was called "head sick"; one that always tipped backward was "tail sick."

When the mill was in motion the matter of stability became a good deal more complicated. For various reasons, including aerodynamic and gyroscopic effects that the early millwrights sensed but did not fully understand, the balance of a windmill is different in operation than at rest. The successful millwright, therefore, needed an accumulation of trial-and-error knowledge that might go back for generations.

The result, even though most everything but the millstones was of wood, was a surprisingly stable and exceedingly durable structure. A post mill built in Lincolnshire, England, in 1509 was still in operation in 1909! And although any storm might leave tragedy in its wake, post mills toppled over less often than their precarious position and top-heavy appearance would seem to promise. In this respect the Williamsburg mill is doubly guarded, being equipped with removable metal braces and buried ground anchors for use in the event a hurricane is predicted. This adaptation, to be sure, is a twentieth-century safety measure, not an eighteenth-century custom.

The problem of balance, and the related difficulty of maneuvering a post mill to face the wind because its whole weight is focused on the one bearing, generally limited such mills to one or two pair of stones. Some English post mills had three pair and a few even four. But these exceptions demonstrate the limitations of the post mill and the reasons for development of its successor, the tower mill.

The purpose in this development was to remove weight from the pivoted upper portion of the mill to solid ground beneath it. In the tower mill, almost the whole mill became a firm structure. Only the cap, holding the sails and their axle, needed to be turned to face the wind. Turning this cap was far easier than turning the whole body of a post mill. Small to start with, tower mills became quite large when mechanical means were developed to adjust sail area. The tallest English tower mills were more than 100 feet high at the hub of the sails, with sweeps that reached out as much as forty feet.

The so-called "smock mill," common in Holland and brought to England probably in the time of James I, is a tower mill whose structure is framed and covered in wood rather than built up of masonry. Examples of this variety of mill can still be seen on Nantucket Island, Cape Cod, Long Island, in Rhode Island, and perhaps elsewhere. The eastern end of Long Island contains more colonial windmills than any other part of the United States today, and all without exception are smock mills.

INDIAN CORN AND COLONIAL MILLS

Both windmill and watermill have been intimately associated with the development of the English colonies in America from their earliest days. This, of course, is not a matter of wonderment since bread was the staff of life then even more than it is today. No doubt all the early settlements had mortars or small hand mills, and in many cases they also employed larger ones powered by animals.

The first settlers at Jamestown in 1607 brought with them full and detailed instructions drawn up in advance by the Virginia Company of London. The 144 men and boys were to be divided into three working groups: one to build a fort, storehouse, church and dwellings; the second to clear land and plant the wheat brought from home; and the third to explore the surrounding countryside in search of the Northwest Passage, mineral riches, or other resources that might return dividends to the company's stockholders.

As it turned out, the planting of grain received less than prime attention. Defense against the Indians was a more pressing demand, and many of the gentlemen settlers were unwilling to soil their hands with menial labor. An exploring

From an etching by James D. Smillie entitled "Old Mills, Coast of Virginia." The original—now in the New York Public Library—was made in 1890, probably on Virginia's Eastern Shore.

party, however, reported that it had observed at the falls of the James River five or six islands "very fitt for the buylding of water milnes thereon."

Several years seem to have passed before any mill was built in Virginia. In 1620 the Company sent word that it considered the construction of watermills of first importance. The next year it specifically instructed the colonists to erect corn mills and bake houses in every borough.

Actually by 1621 the first mill had been put up by Governor Yeardley on his own plantation near the falls of the James River. But it was a windmill, not a watermill, and for at least four years seems to have been the sole facility of its kind in the whole coastal wilderness of North America.

The first mill in the Massachusetts Bay colony, where waterfalls were considerably more frequent and closer to the coast than in tidewater Virginia, was also a windmill, built in 1631. In New Amsterdam the first mill, again a windmill, was erected in 1632. In Virginia by 1649 there were nine mills in operation, four windmills and five water-mills, and the number had grown as fast or faster in other areas.

Exposed coastal areas on Cape Cod, around Newport, Rhode Island, and on the eastern end of Long Island, as well as on the Eastern Shore of Maryland, the Carolinas, and Virginia were found especially well adapted to wind-mills. Massachusetts in particular saw a rapid rise in milling. It was there, under the aegis of John Pearson, "the father of the milling industry," that commercial milling got its early start in America.

At somewhat later stages a similar boom in milling activity took place in New Netherland, New Sweden, and their successor English colonies. For several decades New York held the crown as the wheat-growing, grist-milling, and flour-exporting capital of the New World, only to be superseded about 1700 by Pennsylvania.

WHERE TOBACCO WAS KING

In Maryland and Virginia, where tobacco was the king-sized money crop, grist-milling developed along a some-what different path. Throughout the seventeenth and well into the eighteenth century the tobacco colonists grew wheat and corn for home consumption only. And "home consumption" in most instances meant literally that. The typical plantation, an almost self-sufficient community in many ways, raised wheat enough for the owner's family and sufficient corn to feed the slaves and animals.

In his report of 1724, called *The Present State of Virginia*, Hugh Jones avowed that:

> As for grinding Corn, &c. they have good mills upon the Runs and Creeks; besides Hand-Mills, Wind-Mills, and the Indian invention of pounding Hommony in Mortars burnt in the Stump of a Tree, with a Log, for a Pestle, hanging at the End of a Pole, fix'd like the Pole of a Lave.

Often the planter owned and operated a grist mill for his own use and that of the neighboring small farmers. William Fitzhugh, for example, described his fully equipped layout as including, in 1686, "a good water Grist miln, whose tole I find sufficient to find my own family with wheat & Indian corn for our necessitys and occasions." Fitzhugh, thus, needed to grow no grain of his own to feed his "family," an expression that to him included not only the white indentured servants who lived and worked on the plantation, but also his twenty-nine slaves.

Such a mill represented a considerable investment of capital, and it was this initial cost as much as any other factor that determined the pattern of mill ownership in Virginia. So far as records survive to tell the story, all of the colony's early mills were built on plantations, either by well-to-do colonial officials or by syndicates of neighboring planters. Most of these early mills, if not all of them, were built primarily to grind the owner's produce. Since few but the very largest plantations could keep a mill busy at grinding home-grown grain, most plantation mills also did custom grinding for nearly farmers. A mill formerly owned by John Robinson, speaker of the House of Burgesses, collected enough toll in this fashion to feed a "family" of nearly sixty persons, plus several horses.

It was out of this combination plantation-custom type

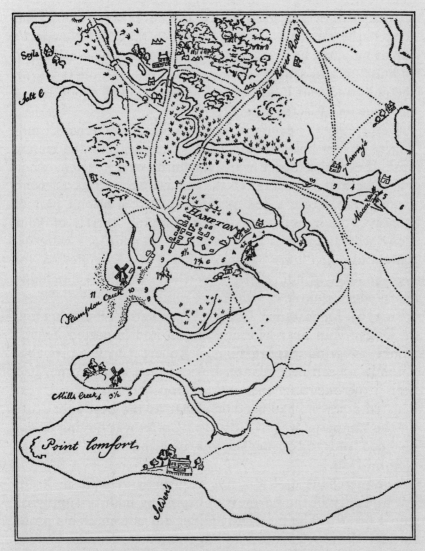

Two tower mills and three post mills can be seen on this map, redrawn from a "Sketch of the East end of the Peninsula Where on is Hampton." On the same peninsula, thirty miles to the northwest, lies Williamsburg. The original map is at the University of Michigan among the papers of Sir Henry Clinton, commander-in-chief of British forces during part of the American Revolution.

of mill that the merchant mill finally developed in Virginia. The first William Byrd was well ahead of his fellow planters in developing milling as a business, though he lagged far behind John Pearson in Massachusetts. In 1685 Byrd had erected two water grist mills at the falls of the James River—the very power site of which Captain Newport's exploring party had remarked. He asked a friend in London to hunt up and send over one or two honest millers to run the mills and sent inquiries to the West Indies about selling the flour he expected to make.

Despite Byrd's example, commercial or "merchant" milling as it was known, gained little headway in the colony until the next century was two-thirds over. Then, when wheat became important to the tobacco growers as a second export crop, quite a few planters added a second pair of stones to their mills and began shipping barrels of flour along with hogsheads of tobacco. The additions were usually buhr or "burr" stones from France, preferred for high-quality grinding because their structure included sharp-edged quartz cavities.

In 1769, for instance, George Washington rebuilt his mill on Dogue Run near Mount Vernon, and imported French stones to grind export flour. Robert Carter, probably Virginia's wealthiest planter-businessman at the time, had been experimenting with other crops on his tobacco-exhausted acres and had fixed on wheat as the best substitute for the "Imperial weed." By 1772 Carter was buying wheat in 8,000 and 10,000 bushel lots to grind in his mill near Nomini Hall.

The merchant mill was not a business venture in itself, but a facility in the business of exporting flour or supplying ship's bread. The merchant miller did not make his profit through the provision of a milling service; he actually bought the grain and processed it on his own account, making a profit or loss on the sale of the product. In many instances the merchant miller not only ground wheat into flour, but also baked the flour into bread for export—particularly in the form of ship's biscuit.

The owner of a custom mill, on the other hand, did not buy and sell grain at all, but found his income as a portion or toll of the grain he milled. The pure custom mill was a rarity in Virginia, however, being limited to a few establishments in and around the towns of Manchester, Petersburg, Norfolk, Alexandria, and Williamsburg.

A mill in Yorktown, too, was presumably of this sort. In 1711 the owners of land on the York River just below Yorktown Creek, deeded a parcel to William Buckner for a windmill, on condition that he "grind for the donors 12 bbls. of Indian corn without toll." A view of Yorktown drawn about 1850 shows an abandoned smock mill—very likely the original one—standing lonely and forlorn on the hill in question.

HOW WAS IT IN WILLIAMSBURG?

Very little information about milling in Williamsburg has survived from colonial days—or at any rate has been unearthed by diligent research. We do know that Williamsburg was considered to offer many choice locations, and that several mills were erected in the town or in its immediate vicinity before the Revolution. As far back as 1699, when the burgesses were thinking of moving the capital from Jamestown to Middle Plantation (as Williamsburg was then named), a student of the College of William and Mary extolled the proposed site in a formal speech, one of several made to an audience that included the governor and his council as well as the burgesses.

The college itself was already located at Middle Plantation, on the ridge between the James and York rivers and with creeks flowing to each. In the words of the student orator, "The neighbourhood of these two brave creeks gives an opportunity of making as many water mills as a good Town can have occasion for, and the highness of the land affords great conveniency for as many Wind mills as can ever be wanting."

All the known watermills lay outside the corporate limits of the city. The nearest, apparently, was a paper mill built by William Parks, founder of the *Virginia Gazette*, and described by him as, "The first Mill of the Kind, that ever was erected in this Colony." It stood about a half-mile south of town on a stream that is known to this day as "Paper Mill Creek."

George Washington, when he married the rich Widow Custis, became the owner of two plantations close to Williamsburg and a water grist mill not three miles from the town. Samuel Coke, the Williamsburg silversmith, owned a grist mill, also water powered, less than one mile away.

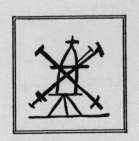

Post mill symbol, re-drawn and enlarged, from the "Frenchman's Map" of Williamsburg, 1782.

What information we have concerning Williamsburg windmills is limited to three fragmentary items. First, in 1723 William Robertson, clerk of the General Assembly, holder of a number of other government offices, lawyer, and land speculator, deeded to John Holloway four lots in Williamsburg, "being the lots whereon the said William Robertson's wind mill stands."

Second, during the Revolution an American soldier who kept a diary of his experiences mentioned being "near the windmill, in Williamsburgh" one night before the siege of Yorktown.

Finally, an unknown French mapmaker, presumed to be in the service of Rochambeau, drew a very careful and complete billeting map of Williamsburg and the buildings in it. On this map appears a representation of a post mill just on the southern edge of the town.

Beyond these three items the story of milling in Williamsburg has to rest on careful deduction, the cross checking of every pertinent fact, the following of every lead, the consultation of every source—in sum, on a mass of research.

For example, evidence as to how the Williamsburg windmills functioned has long since disappeared. Because Robertson's mill was within the city, however, it can be said with reasonable certainty that it operated as a custom mill, not a plantation mill.

MILLING AND THE VIRGINIA PLANTERS

If not much is now known about milling within the confines of Williamsburg itself, a great deal can be related of milling in the larger expanse of the Virginia colony. For despite the late start of merchant milling in the tobacco colonies, the grinding of grain for export had become big business by the time of the Revolution. In 1766 Governor Fauquier noted in a report to the Board of Trade (albeit almost as an afterthought) that the Virginians "daily set up mills to grind their wheat into flour for exportation."

Such merchant mills, as advertisements in the *Virginia Gazette* attest, were usually connected with a plantation. Advertisements for the lease or sale of other farm property repeatedly contained the phrase "convenient to church and mills." A mill scheduled to be built near Robert Carter's on Nomini Creek—too near, he thought—would be the twenty-fourth within twelve miles on the Virginia side of the Potomac, an area that included no towns.

Carter's "new mill," completed in 1773, had a capacity of 25,000 bushels a year and cost him £1,450 in materials and wages. Carter also built in connection with the mill a bake house, the two ovens of which would bake 100 pounds of flour at each heating. And he hired a cooper to "get up 10 good flower caskes per day" at an annual salary of £30.

Carter estimated his total outlay to keep the mill running at £5,000 per year, but the return was correspondingly great. The mill was a success from the start, and the Revolution soon added to its importance and its business. For several months in 1780 the mill worked eighteen hours a day grinding for the state, and Carter received six bushels of corn a day in toll. After 1785, however, Carter found it unprofitable to work the mill himself, and leased it to other operators.

Though on a smaller scale, George Washington also engaged in merchant milling. His Dogue Run mill, formerly a plantation mill only, was rebuilt in 1769 as a merchant mill. He installed a pair of French burr stones to grind the export flour, while a pair of Cologne stones did the country work and ground Washington's own crops. Washington also provided a nearby dwelling house and garden for the miller and his family, who could raise chickens for their own table but never for sale.

"The whole of my Force," Washington wrote in 1774, "is in a manner confind to the growth of Wheat and Manufacturing of it into Flour." Some of the wheat he proposed to sell in London if the price were right, if the freight and commission charges were not too high, and "if our Commerce with Great Britain is kept open (which seems to be a matter of great doubt at present)." He did ship some flour directly to the West Indies but disposed of most of his "superfine flour of the first quality" through merchants in Norfolk and Alexandria.

Washington, like Carter and other Virginia merchant millers, felt the loss of West Indian markets after the Revolution. But he continued his interest in milling into the better times that followed the establishment of a stable national government. Washington, in fact, received one of the first licenses to use the milling improvements invented by the Delaware millwright, Oliver Evans. As late as 1799, the year of his death, he wrote that "As a farmer, Wheat and Flour constitute my principal Concerns."

As a millowner, one of Washington's chief worries seems to have been very much the same as that of William Byrd a century earlier, namely, to obtain the services of an honest and diligent miller to operate his mill. The problem had faced Robert Carter, too, who sent inquiries as far as New Jersey. That other Virginia planter-entrepreneurs faced the same challenge is apparent in many advertisements in the *Virginia Gazette* of the time.

When Washington rebuilt the Dogue Run mill, he was fortunate in hiring William Roberts as miller. Not only was Roberts an honest man in his employer's opinion, but also a highly capable miller. Washington gave him full credit for the fact that flour from Dogue Run commanded top prices in Alexandria and the West Indies markets.

For several years the arrangement was ideal. Then Roberts grew more and more interested in a wheat product other than flour. By 1783 he had become such a drunken sot that the squire of Mount Vernon began seeking a replacement, only to relent when the miller promised to reform his ways. However, this pledge, like its predecessors, soon dissolved in alcohol, and Washington finally fired Roberts.

A substitute, Joseph Davenport by name, was lured from Pennsylvania but turned out to be an inferior miller and as slothful as Roberts had been unreliable. Even so, Washington tolerated him until Davenport's death in 1796. His successor, Callahan, was a competent miller but again far from industrious, and demanded higher wages than the mill could support. In desperation, Washington hunted up Roberts and offered to rehire him on condition of "a solemn and fixed determination to refrain from liquor." This arrangement fell through—perhaps Roberts celebrated too heartily—and the President finally leased the mill to his overseer, James Anderson.

ALL THAT THE LAW ALLOWS

It was said earlier that legal restrictions on milling crossed the Atlantic along with the jolly practitioners of that craft. Indeed, the history of milling in the colonies is fully punctuated by the regular passage or amendment of laws to "rectifie the great abuse of millers," as the first such law in Virginia put it. This first Virginia law appeared as early as 1645 and fixed the allowable toll at a generous one-sixth. Such a law had been passed ten years earlier in the Massachusetts Bay colony.

In neither colony, however, did the law seem to be effective without frequent amendment. The Massachusetts General Court repassed and strengthened its regulation five separate times in thirty years: the Virginia burgesses acted

the same number of times in an even shorter period. A
prohibition against taking excessive toll and the setting of
penalties and fines for violation figured in every revision
of the Virginia law throughout the seventeenth and eighteenth centuries. The basic regulation, passed in 1705, provided:

That all millers shall grind according to turn; and shall
well and sufficiently grind the grain brought to their
mills; and shall take no more for toll or grinding, than
one eighth part of wheat, and one sixth part of Indian corn.

Other phases of governmental interest in grist milling
involved the exercise of the right of eminent domain to provide watermill sites; the requirement that roads be established and maintained leading to mills; that mill dams be
wide enough at the top for a carriage way, include locks for
navigation and fish slopes if necessary, and not be built
above or too close below an existing dam; the inspection of
flour to assure uniformly high quality, free from impuri-
ties; the requirement that millers have and use measures
tested for accuracy; and so on. From such legislation it
will be seen that milling, ostensibly a purely private venture,
partook strongly of the nature of a public utility.

In view of the mill's vital importance to the community,
as revealed in this legislative history, it is no surprise to
learn that the miller was considered essential too. Along
with certain officials of the colony, clergymen, plantation
overseers, the jailer, schoolmasters, and some other groups
deemed necessary to orderly civil life, millers were exempt
from service in the militia. Furthermore, since militia
musters were often occasions of prolonged revelry, any
miller who "shall presume to appear at any muster" was
to be fined one hundred pounds of tobacco or be "tied Neck
and Heels" for up to twenty minutes. Only when the need for
foot soldiers became all-consuming in 1780 was the militia
exemption lifted. Until then the miller was expected, and
obliged, to keep his nose to the millstone.

"Militia musters were often occasions of prolonged revelry. . . ." Adapted from an engraving by the English painter and caricaturist of the eighteenth century, William Hogarth.

The miller, thus, seems to have had a split personality—at least in the public mind. On the one hand he possessed an ancient reputation for dishonesty that called for repeated legislative curbs and punishments. On the other hand, he was so indispensable to community welfare that the law got after him if he took a day off for public carousing as other men did.

At least since Greek and Roman times the miller, who performed a task once relegated to women and slaves, was traditionally held in low esteem by reason of his calling. Yet some colonial millers were respected and influential men, and sometimes men of substance. John Jenny built the windmill in Plymouth, Massachusetts, in 1636 and was chosen by his fellow townsmen to represent them in the General Court. Two years later he was indicted for failing to grind his neighbor's grain well and seasonably. The nearby town of Rehoboth similarly elected its miller to the General Court. But he refused to leave his mill in order to serve as deputy and was fined!

In colonial Virginia the social position of the miller was less subject to violent fluctuation than would seem to have been the case in New England. In fact the Virginia miller was uniformly a man of low estate, far inferior to the owner-operator of a mill in a New England town, and outranked also by the sturdy bourgeois millers of the middle colonies. Those who worked for wages enjoyed few privileges, while the many Virginia millers who were either Negro slaves or white indentured servants had little social standing. The records contain a goodly number of references to runaway millers who were indentured servants, convict servants, or slaves; if any Virginia miller in colonial times rose to a position of importance, no record has yet been found.

THE SAID WILLIAM ROBERTSON'S WIND MILL...

Most windmills belonged to one or other of the two basic categories: post mill or tower mill. The truth of this general rule is easily proved by the exception—the composite mill that belonged to neither group, but was in effect a post mill set on top of a tower. Another truism—perhaps without any exception—states that since every mill was custom designed and made by hand, no two were exactly alike.

Every English and colonial millwright had his favorite tricks of design and construction, and often the mills of a region showed a family resemblance that distinguished them from the mills of the neighboring shire or colony. In addition, improvements and refinements were developed from time to time and gradually put into use. Many of them, like the grain elevating machinery invented by Oliver Evans of Delaware and the adjustable lattice sails of the English inventor Sir William Cubbitt, came long after 1716-23, the years during which Robertson's original mill was built.

Clearly, this is not the place in which to describe the almost infinite variety in structure and operation of windmills in different places and at succeeding stages in the perfection of the mill as a machine. A list of a few easily procured books appears on page 32 for those who wish to pursue the subject further. This little pamphlet must be limited to a description of the reconstructed mill in Williamsburg and how it operates.

First, a word needs to be said about the men behind the mill, Edward P. Hamilton, former director of historic Fort Ticonderoga, New York, and Rex Wailes, then of London, who were Colonial Williamsburg's consultants in the careful process of design and construction. Mr. Wailes, a world authority on mills and milling machinery, furnished information on all phases of the use of windmills in England, and in particular provided measured drawings of a seventeenth-century mill still standing at Bourn in Cambridgeshire. Mr. Hamilton carried on from there. By vocation an investment counselor (retired) and by avocation a collector of watch and clock mechanisms, an authority on windmills in America, and a skilled model builder, he transformed the drawings into miniature reality, creating a perfect working model in which every part performs the function assigned to its larger counterpart in the mill itself.

William Robertson, who sold to John Holloway in 1723 four lots near the Palace "whereon the said William Robertson's wind mill stands," was not a miller. Neither was Holloway. The records left by both men provide no clue as to the exact appearance of the mill in question. In fact, the deed to the land itself did not give any more precise location for the mill.

Reconstruction of the mill, therefore, had to depend on answers to two questions: Just where did Robertson's windmill originally stand? and What kind of a mill was it?

By eliminating every other possibility, the site of the lots was established with certainty at the corner of North

Robertson's Windmill, a faithfully authenticated post mill, stands in Williamsburg today on a spot near the Governor's Palace where its predecessor is believed to have stood about 1720.

England and Scotland streets. By elimination again, the spot where the mill must have stood was placed at or near the present site—simply because other buildings were known to occupy various other locations in the tract. Thorough archeological excavation of the whole area, however, disclosed no corroborative evidence to show the precise location of the mill.

At the same time, the absence of such evidence was itself almost conclusive evidence that the mill was a post mill such as were common in tidewater Virginia at that time. A tower mill, which was the second variety often built hereabouts in colonial days, would have had foundations in the ground. Traces of these would have been revealed by digging at the site. Since none of the foundations excavated in the four lots was suitable to the underpinning of a tower mill—ergo the mill must have been supported above ground on a wooden post.

And so it was rebuilt: a small post mill, simple in design and operation, with but a single pair of millstones. The tail pole, extending to the rear, ends in a wagon-wheel support that eases the miller's task of turning the mill by hand to face the wind. Other millers sometimes had shoulder yokes attached to the tail pole, or pulled it around with the help of a small winch anchored to one of a circle of posts.

The stairway from the ground up to the body of the mill, called the "ladder," is hinged at the upper end so that when the mill is to be turned, a lever raises the foot of the ladder off the ground. When the mill has been positioned to face the wind, the end of the ladder is lowered to the ground again where it helps hold the mill against further turning.

Beneath the body of the mill are the heavy timbers on which it rests: the horizontal "crosstrees," the sloping "quarterbars" and the great post, all hand hewn of well seasoned oak. The tree from which the post is hewn was itself young when Williamsburg first became the colonial

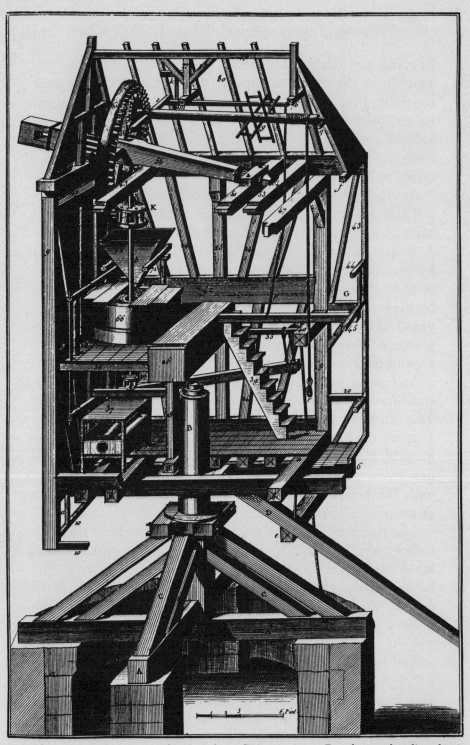

An eighteenth-century cutaway drawing, from Diderot's great French encyclopedia, showing the structure and mechanism of a post mill. The artist has "lifted" the body of the mill a few feet off the post to reveal the pivot bearing on which the massive crowntree would in fact rest. Note also the power takeoff from the brake wheel for the sack hoist.

capital. A count of its annual rings shows it to have been a sapling in 1675, which makes it one of the town's oldest "antiques."

Mounting the ladder and looking into the mill's lower or "meal floor," we see that the post ends in a wrought-iron bearing set into another beam of impressive size that "crosses" the post like the top of a giant "T." This beam is the "crowntree," on which the framework of the mill body is built. Below the lower floor is another bearing or loose collar around the post to keep the body steady on its perch.

The prinary machinery of the mill consists of sails, stones, and the necessary shafts and gearwheels to transmit power from the sails to the stones. In addition, there are devices for braking the sails, for hoisting bags of grain from the ground to the upper or "stone floor," for feeding grain at the proper speed to the stones, for warning the miller when the supply of grain in the hopper is getting low, for adjusting the distance between stones, and for separating the meal from the bran.

The sails of the Williamsburg mill are of the early pattern in which the backbone of each sail frame extends along the centerline of the sail. That is, the area of sail ahead of the backbone—in the direction of turning—is the same as the area following it. Incidentally, windmill sails usually turn counterclockwise (viewed from in front of the mill).

The sailcloths themselves are handmade, of imported Scottish linen. To meet changes of weather they are furled by hand, each arm in turn being stopped at the lowest position while the miller unties the outward corners and twists the long strip into a more or less tight roll. For this purpose he can set and release the brake from the ground, using the rope that hangs out the side of the mill.

Whereas each sailcloth of Robertson's windmill must be wholly furled or not trimmed at all, the sails of many early mills could be partially reefed. The four degrees of reefing were known as: full sail, first reef, dagger point, and sword point. Trimming the sails was a difficult and sometimes dangerous task, for a sudden storm with sleet and shifting gusts of wind could make the job almost impossible at the very moment that it had to be quickly accomplished. A miller caught with his sails up in such a storm might suffer what was known as "tail winding" if the wind veered faster than he could work. In this event he might be lucky to get off with nothing worse than having the sail-cloth stripped from the frame.

A hurricane could do more serious damage—and might overturn the whole mill—no matter which way it faced. For the wind was the miller's master as well as his servant, an evil genius that he feared as well as a heavenly blessing for which he prayed. Without wind the mill stood idle and the miller earned nothing. When the wind arose the miller must heed its call to work, whether in the middle of a meal or in the middle of the night. And always, he must keep a weather eye on the horizon for signs of too much wind.

Fire and lightning were other great perils to every windmill. If the hopper ran empty of grain, the friction of the stones rubbing against one another could generate enough heat for combustion. So could the friction of the brake if it were used continuously in a strong wind. In either event, a building made entirely of wood and open to every breeze burned readily, and more windmills probably fell victim to fire and storm than to old age. Similarly, its height and exposed position made the windmill an attractive target for lightning.

The sail wheel of the Williamsburg mill has a diameter of fifty-two feet—about average for a small post mill of this type. In a wind of about twenty miles an hour, which seems to be generally best for windmill operations, it will turn at about twenty revolutions per minute. This apparently slow and majestic rate is deceptive; for at twenty rpm the tips of the sails travel at a linear speed of 3,266 feet a minute or about thirty-seven miles an hour! An operating windmill is something to stay well clear of, as Don Quixote and uncounted innocently grazing cows and sheep have discovered to their sorrow.

The four arms are fixed into the hub of the massive "windshaft." This is the main horizontal axle that brings the power of the turning sails into the mill. Actually, it is not exactly horizontal but slants at a ten-degree angle. Since the wind has always been thought to descend from heaven, this ancient arrangement may at one time have been intended to let the sail wheel "look up" a bit into the wind. It has, in any case, both structural and aerodynamic advantages.

Just inside the front wall of the mill the windshaft rests in a metal bearing, and next to it is the large gear wheel, known as the "brake wheel," that can be seen when one looks up through the trap door from the mill's little back platform. The platform, incidentally, was not unusual in post mills, though it was often a convenience added after the mill was built and on which the miller could enjoy a pipe and a moment of repose while the mill ground merrily away.

The brake wheel, a little more than seven feet in diameter, is firmly fixed to the windshaft and turns at the same speed as the sails. Its fifty-one hickory gear teeth and eighteen other major pieces, plus at least as many wooden pegs to hold the pieces together, were all carefully shaped and fitted together by hand. Around its outer edge is the brake band —of bent hickory—that can stop the sails and hold them at any position when the long, heavy brake lever is lowered. As an emergency aid in slowing down the sails in a strong wind, the stones can be choked with grain, thus making them harder to turn. The mill can also be operated with partially furled sails.

The stones themselves consist of the lower or "bed stone," which does not move, and the upper or "runner stone," which turns at a little more than five times the rate of the sails. Its normal speed for best results is 108 revolutions per minute, at which speed it produces five to ten pounds of meal per minute.

*Another engraving from Diderot's encyclopedia show-
ing the passage of grain from hopper to millstones, and
its reappearance as meal. Notice the arrangement to set
the bell ringing when the hopper is about empty and
the weighted lever at the left by which the miller can
control the distance between the millstones.*

The cogs of the brake wheel stay in constant mesh with
the staves of a sort of wooden-bird-cage gear called a
"wallower." This is fixed to the upper end of the wrought-
iron vertical drive shaft, whose lower end stands in the
center or "eye" of the runner stone and turns it.

This simple drive mechanism is complicated by the fact
that the runner stone must be held suspended above the
bed stone. Ideally, the faces of the two stones—however
close together—should never touch. To accomplish this,
the runner stone is balanced to turn freely on the point of
a spindle that comes up from below through the eye of the
bed stone. The spindle, in turn, rests at its lower end in a
pivot bearing that can be raised or lowered very slightly by a
series of levers.

By this means the miller can adjust the distance between stones according to the kind of grain he is grinding and also according to the speed at which his mill is operating. In a variable wind, for instance, he may have to make the adjustment continuously as the feel of the meal coming from the stone indicates.

The faces of both stones must be "dressed" periodically—perhaps once every ten days when the mill is in constant operation. For this the runner stone must be removed and turned over so that the stone dresser, with his pickaxe-like "millbill" can operate on both faces. The dresser deepens the furrows, if necessary, and roughens the "lands" between the furrows toward the outer edges of the stones.

When the stones are in good condition, the grain kernels are opened out near the eye of the stones, gradually reduced in the middle area, and the bran scraped and cleaned in the outer one-third of the face. The condition of the bran is the best index of the stone dressing, while the feel of the meal tells the miller whether his mill is running at the best speed and the stones are set at the proper distance apart.

French burr stones produced the best quality flour in colonial mills, and most Virginia mills that ground for export seem to have had a pair of them. Cologne, or "cullin," stones from the Rhine were somewhat less choice, and the same mills often had a pair of these for country work, especially for grinding corn. Stones quarried in Virginia, Pennsylvania, New York, and elsewhere were also used. The pair in the Williamsburg mill are of quartz-bearing granite quarried in Rowan County, North Carolina. They are four and one-half feet in diameter. The bed stone is seven inches thick and the runner stone ten inches. Together they weigh more than two tons.

Although colonial millers ground all varieties of grain between the same set of stones—making only the necessary adjustment in distance between the stones—Robertson's windmill today processes only corn. But whatever the grain, the difference of a newspaper's thickness makes the

The stone dresser often steadied his forearm on a sack of meal as he worked. His hands were constantly bombarded with bits of stone and slivers of metal from the point of his pick or "millbill." Some of these slivers became imbedded under the skin, and an itinerant stone dresser looking for work could prove his experience by "showing his metal," i.e., the backs of his hands.

difference between a good grind and a poor one. So for all its clumsiness in appearance and lumbering clatter in operation, the windmill is an extremely precise mechanism at the point where precision counts.

The grain, raised bag by bag through trapdoors on a rope sack hoist, is poured into a hopper above the stones, which are themselves entirely enclosed in an octagonal wooden box called the "stone casing" or "vat." Suspended from the bottom of the hopper, an inclined trough or "shoe" carries the grain to the central opening of the runner stone. The turning of the drive shaft constantly joggles the shoe, and by raising or lowering one end of it the miller can regulate the flow of grain to the stones.

The grain is ground as it works its way outward between the stones. At the outer edge of the bed stone the meal falls into the narrow space between the stone and the casing.

There, air currents moving with the turning stone continually sweep the meal around to an opening in the floor of the casing.

From this opening a chute leads downward to the sifting device on the lower floor. The sifter, on the left side of the mill as you look in from the platform, separates the bran from the meal. The miller can bag each product almost automatically, since the sifter, too, is constantly shaken by an ingenious connection to the mill's driving mechanism.

The entire mill, thus, is about as simple a machine as one could devise. The moving parts are few and could be called primitively clumsy in comparison to some of the sleek masterpieces of modern industrial design. The miller can make only three operating adjustments: in the sail area presented to the wind, in the rate of grain fed to the stones, and in the distance between the stones. He has no control over wind speed, cannot shift the gear ratio of the mill— or even shift the stones out of gear—and finally, cannot determine his own hours of work and rest.

Perhaps the reader will be inclined to marvel a bit that the product of such a mill is so good. And perhaps he will look with some compassion on a man who is so completely a slave to the elements. If some of the miller's fellows have been disposed—on occasion—to take just a trifle more toll than law and custom allow, well, let him who is without fault cast the first stone. Millstone, that is.

SUGGESTIONS FOR FURTHER READING

PAUL G. E. CLEMENS, *The Atlantic Economy and Colonial Maryland's Eastern Shore: From Tobacco to Grain.* Ithaca, N. Y.: Cornell University Press, 1980.

OLIVER EVANS, *The Young Mill-Wright and Miller's Guide.* New York: Arno Press, 1972 (reprint of 13th edition, 1850).

STANLEY FREESE, *Windmills and Millwrighting,* new edition. Newton Abbot, England: David and Charles, 1971.

D. W. GARBER, *Waterwheels and Millstones: A History of Ohio Gristmills and Milling.* Columbus: Ohio Historical Society, 1970.

ROBERT J. HEFNER, *Windmills of Long Island,* ed. T. Allan Comp. Setauket, L. I., N. Y.: W. W. Norton, 1983.

CHARLES HOWELL AND ALLAN KELLER, *The Mill at Philipsburg Manor, Upper Mills and a Brief History of Milling.* Tarrytown, N. Y.: Sleepy Hollow Restorations, 1977.

WILLIAM B. LLOYD, *Millwork: Principles and Practices, Manufacture, Distribution, Use.* Chicago: Cahners Publishing Co., 1966.

BARTON MCGUIRE, comp., *Mill Primer: A Glossary of Milling.* Water Mill, L. I., N. Y.: Water Mill Museum, 1977.

LOUIS MORTON, *Robert Carter of Nomini Hall: A Virginia Tobacco Planter of the Eighteenth Century.* Charlottesville: University Press of Virginia, 1964.

JON A. SASS, *The Versatile Millstone Workhorse of Many Industries.* Knoxville, Tenn.: Society for the Preservation of Old Mills, 1984.

JOHN STORCK AND WALTER DORWIN TEAGUE, *Flour for Man's Bread: A History of Milling.* Minneapolis: University of Minnesota Press, 1952.

VOLTA TORREY, *Wind-Catchers: American Windmills of Yesterday and Tomorrow.* Brattleboro, Vt.: S. Greene Press, 1976.

REX WAILES, *The English Windmill.* With drawings by Vincent Lines. London: Routledge and Kegan Paul, 1954.

HARRY B. WEISS AND ROBERT V. SIM, *The Early Grist and Flouring Mills of New Jersey.* Trenton: New Jersey Agricultural Society, 1956.

The Miller in Eighteenth-Century Virginia was first published in 1958 and was reprinted in 1966, 1973, 1978, 1988, 1990, 1997, and 1998. Written by Thomas K. Ford, late editor of Colonial Williamsburg publications, it is based largely on an unpublished monograph by Horace J. Sheely, formerly of the Research Department.

ISBN 0-910412-19-7

90000

9 780910 412193